Working Dogs

Working Dogs

Audrey and Edward Hart

B. T. Batsford Ltd.   London

First published 1984

ISBN 0 7134 4781 8

Typeset by Tek-Art Ltd Kent
and printed in Spain
for the publishers
B. T. Batsford Ltd.
4 Fitzhardinge Street
London W1H 0AH

Frontispiece *Training a bloodhound for a search and rescue team in California*

ACKNOWLEDGEMENTS

We would like to thank Maureen Nixon, Val O'Keefe and Geoff Hargreaves, who motored many miles to help us in the search for gundog photographs. We are also grateful to Cumbria Constabulary, Penrith, the members of Leamington Dog Training Club, June Hardy who typed the manuscript and Peter and Audrey Foster whose hospitality and help with photographs of Flat-Coats was invaluable.

Acknowledgement for the photographs is due as follows: Mike Amps, Search and Rescue Dog Association (England): page 40; Sergeant John Duff, North Eastern Counties Constabulary: Frontispiece, pages 49, 50; John Grey, Search and Rescue Dog Association (England): page 41; Guide Dogs for the Blind Association: pages 43, 44/45; Audrey Hart: pages 4, 11, 12, 13, 14, 15, 16/17, 18, 19, 20, 21, 22, 27, 32/33, 34, 35, 36, 47, 52/53, 54, 58, 59, 60; Edward Hart: pages 6, 10; Marc Henrie A.S.C.: pages 8/9, 24/25, 26, 39; ICI Dulux: pages 56/57; Panther Photographic: page 48; Public Information Department, New Scotland Yard: pages 37, 38; Alan Thompson: pages 29, 30/31.

CONTENTS

Introduction

The well-trained working dog is an extension of its handler. It has been used by man since animals were first domesticated, and has become so much part of his being that, for many, life without a dog would be unthinkable.

Just as breeds of dogs vary enormously in size, type and temperament, so do the jobs which they undertake. In this book we are concerned with dogs that work – as sheepdogs, gundogs, guide dogs, police dogs etc. – as distinct from those which provide invaluable companionship as pets or those which are on show as exhibition animals.

Classifications are elastic. Some of the best gundogs ever known have been Border Collies, a breed more usually associated with sheep herding. One of the most indefinable mongrels I ever saw had an excellent nose. He would push up game at just the right distance in front of the guns on a rough shoot.

Constant association with its owner brings out the best in the working dog. One sheepdog regularly rides on the tractor transport box, and as his farmer owner drives along to the sheep's feeding point, he hops off, lopes up the hillside, and brings along any stragglers. There is no discernible command.

The best working dogs love their jobs. No amount of thrashing is going to send a Lakeland Terrier down a fox earth. He goes of his own accord; the difficulty is in holding him back. When a terrier has been rescued after spending several days down a hole without food or water, the first essential is to take a grip firm enough to ensure that he does not immediately go to ground again.

It is quite obvious that the sheepdogs seen on the BBC television series *One Man and His Dog* adore their masters. Their keenness and instincts have been channelled, not beaten, towards the desired goal. Foxhounds crowding the huntsman at a meet gaze up at him with the same soulful eyes as a Spaniel's towards his keeper.

Mountain and sea rescues are undertaken by dogs, as are searches following the devastations of air raids or earthquakes. The dog is schooled by reward, often through its stomach. The happiest of all dogs are those which work at the task for which they were bred and trained; they ask no more than to be allowed to work, and to receive praise, affection, an adequate meal and a warm bed.

Hunting on foot in the English Lake District is not just a pastime – it is essential to the sheep farming industry

Border Collie 'ratting' in stubble

1 Farm Dogs

Man was not slow in using dogs to help him domesticate other animals. At first the dogs were guards; they still are, in some countries. Hungarian Komondor, Maremma, and Pyrenean Mountain Dogs are among those being used to protect people and property.

At New Hampshire Agricultural College, livestock guard dogs are being developed to stay with the flocks, and travel with them day and night. The dogs act as a warning to intruders, placing themselves between undesirable people and the animals. But in the case of predators, the guards will attack with powerful teeth and jaws.

Packs of house dogs are a menace in the southern USA, and here guard dogs are proving effective. There is a useful role for such dogs in Britain's urban fringe, where stray dogs and trespassers are a constant worry. In the early days, after predators such as wolves had disappeared, the guard dogs became family companions. Now they are reverting to their former role.

The shepherd's and his dog's combined skills have

Border Collie and sheep in evening light

been brought to a wide and appreciative audience through the successful BBC 2 series *One Man and His Dog*. The trials are designed to reflect working needs – gathering a flock of sheep widely scattered over a large area, fetching them straight to the handler, then driving them away from him.

The 'shed' is the splitting off of unwanted sheep in the open, far from handling pens. At trials it is done by shedding sheep with a particular mark; on the open hill the shepherd sheds off strays from another flock, recognizing them as readily as a suburban man recognizes his neighbours.

The Border Collie

Before the Border Collie was developed, men and youths ran over the hills of Wales to gather the sheep. In those days the uplands were more heavily populated, but as the nineteenth century developed, more sheep were worked by fewer men.

The Border Collie was developed in the Scottish Border country from working sheepdogs from all over Britain. The Scotch Collie, which has now become the 'Lassie' or Rough Collie, is a show dog only. Early breed histories show the point of divergence.

By the mid-nineteenth century a sufficient number of sheepdogs had been sent to Wales from the Scottish Borders to affect local strains noticeably. The first-ever sheepdog trial was staged at Rhiwlas, near Bala, in North Wales, by R. J. Lloyd Price, a great gundog man. His grandson and great grandson are still farming the same estate, and the Centenary Trials were held there in 1973.

Sheep are controlled by the Border Collie's 'eye', which induces a kind of hypnosis. A 'strong-eyed' dog commands its sheep from a distance, and the factor cannot be assessed except by work on sheep. It has no connection with other physical characteristics. This 'eye' is a refinement of ancient hunting instincts. In extreme it is a nuisance as the dog simply

Candidate for the world's most intelligent and versatile breed, the Border Collie is used on farms, as a gundog, as a drug detector, as a guide dog and on the stage; they are also widely favoured in Obedience and working trials. A long working life is an added characteristic – this bitch is 12 years old

Not all Border Collies are black and white. In the centre is a red merle, the first one ever registered, and behind her is her sire, a fine blue merle. The mother was black and white, carrying a brown gene

Opposite Penning at an international trial. Dogs tend to react differently when away from home

lies down and stares at the sheep. Such 'clapping' dogs are less common now than they were 30 years ago.

Many of the best modern Border Collies are registered with the International Sheep Dog Society. While the charge that a trial winner cannot do a day's hill work has often been disproved, some people believe that certain strains are now becoming too highly bred. Temperament tends to be less placid than in the past; the dog kept at heel while its partner is working is more on edge, less content to wait its turn.

The Border Collie is now recognized by the Kennel Club, following pressure from show enthusiasts and Obedience handlers. But the true working strains are firmly based on the hills of Britain, and indeed in most parts of the world where sheep are kept. One-third of the earth's thousand million sheep are worked by Border Collies.

The Sheepdog's Year

On sheep farms, the Border Collies are absolutely essential to the shepherds. Their year begins in autumn, when the rams are turned out with the ewes. 'Tup time' is as vital as lambing time, for it is the shepherd's job to see that all ewes under his care are mated, and by the proper sire.

There may be a stray (unfenced hill) of several square miles, with different groups of sheep on it. For a five-week period, each group of ewes, with distinguishing colour marks, must be gathered round the appropriate ram, a series of long walks for the shepherd and even greater distances for his dogs.

Without dogs, it would be impossible to fit the routine into the short November days. Neither sheep nor cattle respond to a man as they do to a dog, and few things are more exasperating than finding oneself without a dog, and with straying cattle or sheep to

A red-and-white Border Collie holding up Welsh Mountain sheep

retrieve. The tupping round affords a fine opportunity to train young dogs. Sheep respond, are not heavy in-lamb, and have no lambs at foot to arouse their protective instincts.

Winter work depends upon the weather. If storms break, the dogs are used to find sheep buried in snow drifts. Ability in this direction seems unconnected with aptitude at other work. Collies save thousands of sheep year after year in snow. Spring brings daffodils, lambs and work. For the Border Collies and shepherds, dawn to dusk entails at least a 16-hour day, and the worse the weather, the more important shepherding becomes. As each ewe lambs, she is taken from the lambing enclosure (which may be a quarter square mile in extent) and the fiercely protective ewes cause unremitting work for the dogs.

Some ewes, especially younger ones, may not be too keen on their new offspring. A Collie is commanded to approach the offending ewe, and as the dog 'sets' a few feet away, the ewe's mothering instincts are aroused. A good lambing dog is invaluable. It needs courage and sense, for too much force must not be used.

After lambing, the expanded flock is shepherded twice a day, providing more work for the Collies. As the lambs grow, a daily round suffices but the dogs are constantly active, joining up lambs that have lost their dams, retrieving strays through a hole in the hedge, or pushing the sheep into a corner and allowing them to string out of it for counting.

Sheep are gathered into yards for injections, worm doses, shearing and dipping. The lambs are weaned from their mothers, and later bundled into a wagon for the busy sale. Even the pleasant early autumn period entails a great deal of gathering, as breeding sheep for the next year are selected. Then comes 'tup' time once more.

Other Farm Dogs

Cattle dogs include the Corgi, whose low-slung carriage enables it to run in, nip a bullock's heel, drop flat, and so escape the lashing hoof passing over its head. This instant drop is still seen.

The Welsh Heeler is perpetuated in the Cardigan Corgie, and another survivor is the Lancashire or Ormskirk Heeler; this breed was only recognized by the Kennel Club in 1981, when about 200 were thought to exist.

2 Gundogs

Some years after World War II, the long-established classification of British gundogs was affected by the introduction of Continental breeds. These are classified by the Kennel Club as Hunt-point Retrievers. They hunt and flush game like a Spaniel, point or set it, and then retrieve.

Until that time, functions had been more specialized. Setters and Pointers ranged and found the game, indicating its presence by their reactions. Spaniels flushed birds in front of, but within range of, the guns, while Retrievers brought them to hand.

The shooting man seeks in a dog those attributes in which he is himself deficient. By far the most important is scenting power. The human nose is very rudimentary compared with a dog's, and in gundog breeds scent has been highly developed.

A party of guns may walk in line 20 or 30 yards apart through heather, deep bracken or scrub woodland, and leave behind them any number of close-sitting birds. The same party with well-bred

A German Shorthaired Pointer retrieving dead game from water

A Flat-Coat happily crossing a stream while carrying a dummy

A pair of Chesapeake Bay Retrievers

and highly trained gundogs can be sure that virtually all game has been located.

Field Trials are organized to improve the working abilities of a dog which accompanies its master on the shooting field. The quarry may vary from grouse on high moors to wildfowl on an estuary, and the gun may be either walking or having birds driven to him. Whatever the class of shooting, the demands on the dog are much the same.

Dogs are assessed for their ability to find, for their steadiness, reactions to signals, obedience and agility. Some are also required to retrieve from water.

Pointers and Setters

These are used for much the same purpose, and training for both is similar. Important among early obedience lessons is the 'down' or 'drop': the young dog is taught to lie flat on the ground by having a hand pressed on his back; then the handler walks further and further away until he signals the dog to move.

The young dog is next taught to quarter the ground, and to stay within range; its natural instinct is to point any quarry, usually with one forepaw raised. Two dogs work together, and the one 'backs the point' of the other. A pair of well-bred pointers finding game provides one of the greatest thrills in the dog world.

Pointers and setters can scent and point game from more than 40 yards away. The two types reached a peak of popularity in early Victorian times, before the introduction of the reaper and self-binder. When corn was cut by hand, much longer stubbles were left, and the whole of a stubble field had to be worked. When only short stubbles are left any game makes for the hedgerows or the next field.

The Setter is claimed to be the oldest breed of gundog, and the oldest type of Setter is probably the English. It is lower on the leg than the Irish or the Gordon, for it was originally worked over grassland, roots and stubble rather than through deep heather or bog. The amount of feather on the tail of this affectionate breed is a distinguishing mark, and its colours are black-and-white, lemon-and-white, liver and white mottled or tri-colour.

'As red as an Irish Setter' is a universally understood description; yet in the mid-nineteenth century, there was a fair amount of white in the breed. In the 1874 Dublin show, two classes were held for white-and-red Irish Setters, and only one for reds. 'Rich golden chestnut' is the official colour of this breed, which excels in keenness and as a game finder. Endurance and wide-ranging capabilities are other characteristics, so ample work or exercise is needed by these beautiful dogs.

The Gordon Setter is of ancient lineage, black-and-tan in colour, and possesses immense stamina. Its name was conferred on it by the Dukes of Gordon at Gordon Castle, and a priced sale list of 1836 mentions nine Setters, though only one is described as black-and-tan. A colour plate by Sydenham Edwards of 1805 shows three Setters, one of which is undeniably black-and-tan. Captain L. C. R. Cameron in *The Lonsdale Library* series of books puts forward evidence that Gordon Setters were known as such in the Highlands as early as 1746.

Spaniels may be taught to retrieve, and this one has a lovely soft mouth

Retrievers

When walking up game or going over a rough shoot, the Retriever may seek birds or rabbits in front of the guns. Its more specialized job is to wait at the handler's side during a drive, and retrieve on command any birds that are brought down. For this a 'soft mouth' is essential. The object is *not* to half-pluck a cock pheasant on bringing it to the gun!

The five varieties of Retriever are the Flat-Coated, the Curly-Coated, the Golden, the Labrador, and the Chesapeake Bay.

The Flat-Coated Retriever is medium-sized, active and handsome with a dense, harsh coat of black or liver, that enables it to withstand prickly, rough cover. Originally termed Wavy-Coats, this branch of the family was first shown as a defined breed in 1873. Its coat is dense, harsh and about 5 cm (2 in.) long, and lies flat to the body. It has a feathery tail, though less pronounced than that of the Golden Retriever.

Before World War I the Flat-Coat was the gamekeeper's favourite but many of these men joined up early, and did not return. This setback had its effect, but in recent years a number of enthusiastic and capable breeders have put the Flat-Coat to the fore once more.

The Curly-Coated Retriever is especially prized for retrieving in water, and is thus a favourite of wildfowlers. Black or liver in colour, it is a kind and unusual dog, slightly heavier than the Flat-Coat or the Golden.

Like most gundog varieties, the Golden Retriever is bred larger than average for the showbench and smaller for field work. Its long tail hairs set it off beautifully, but tend to collect burrs. An active, powerful and kindly dog, it was only kept by a few

A rare German Wire-haired Pointer backing the point of her Smooth-haired partner

large-scale landowners before the showbench brought it wider popularity.

Labradors are a very popular breed with gamekeepers. Their potential is such that skilled professional training brings them to a very high standard. Their coats of black, yellow or liver are easily cared for. Labradors are used as 'sniffer' dogs (page 38) and also as guide dogs for the blind (page 42).

From North America comes the Chesapeake Bay with its short, thick coat varying from dark brown to straw-coloured. It is rugged and powerfully built, with amber or yellow eyes, and makes an excellent guard dog. In 1982 there were 63 in Britain.

Spaniels

Of all gundogs, these are the best known by the non-shooting fraternity. Boisterous and almost untirable, keepers say that a Spaniel 'requires ten days' shooting every week.' It is used both for flushing game and for retrieving.

The Cocker Spaniel is the most popular. Despite its energy it makes an excellent family companion, although it needs regular grooming. A choice of 14 colours is available.

Heaviest and slowest of the Spaniels, the Clumber can make a splendid retriever. The preferred colour is white with lemon markings. With stately dignity and measured tread, the Clumber is very different from the impetuous Cocker or Springer.

A Springer Spaniel was mentioned by Gervase Markham in 1655, and ever since has been the helpmate of rough shooters and wildfowlers. It was once used to find and spring game for falconry or coursing. The colours of the English Springer are liver-and-white or black-and-white, with tan markings allowed.

The Field Spaniel has a coat entirely of one colour. Originally black, it may now be liver, golden/liver or mahogany red, though possibly with tan over the eyes. It is level-headed, intelligent and docile.

The largest of the Spaniels is the Irish Water Spaniel, with its rich liver-puce coat of dense curls, quite impervious to bad weather. In his classic *Letters to Young Shooters,* Sir Ralph Payne-Gallwey wrote:

> To the shore shooter or one who tramps the fens and bogs, we have found no dog to equal the brown Irish water spaniel. He is happier in the water than out of it, and from his small size can follow a teal or duck almost up a drainpipe. He is a very keen sporter, an affectionate companion, and surely no animal in the world is more intelligent.

The classic pose: Weimaraner demonstrating grace and beauty on the field

The Sussex Spaniel has a liver-brown coat and moves with a definite roll, while the Welsh Springer is always a rich red and white. The latter is a distinct variety, and is of ancient lineage. In his native land he is also know as a 'Starter'.

Hunt-Point Retrievers

Four of the six Continental gundog breeds used in Britain are of German origin. The German Wire-haired Pointer has a strong coat that needs regular grooming, and it delights in a lot of attention and exercise. It has strong guarding instincts. Colours are solid liver, liver-and-white and black-and-white.

The German Shorthaired Pointer is the more popular of the two, and also needs companionship and work to fulfil its great potential. It is a natural retriever with a good nose, and weighs 25-32 kg (55-70 lb). Colours are solid liver, liver-and-white spotted, liver-and-white ticked and black-and-white.

Easily identifiable is the Weimaraner, with its light grey coat. Its tail is usually docked. Another triple-purpose breed on the shooting field, it is also used as a police dog. It is an alert dog with great stamina.

A more recent introduction is the Large Munsterlander. In appearance it has more of the Spaniel than the Pointer, and it is larger than the Springer Spaniel. The Large Munsterlander's colouring is black and white, often attractively mottled. It is a fine water dog.

The Britanny Spaniel has been known in France for centuries, but only a few are in Britain. Liver-and-white or orange-and-white, its size and manner of working suggest a setter. The breed was introduced to the USA in 1931, and became popular as a show and gundog.

Better known is the Hungarian Vizsla, a lean, close-coated dog of copper red colouring. It is easy to train, is good in water and makes a useful retriever. It stands to a height of 57-65 cm (22½-25½ in.) at the shoulder.

A Munsterlander: excellent in water

3 Hounds, Terriers and Racing Dogs

Hounds

'If you've never seen hounds break cover on a fine winter's morning, you've never lived!' This Yorkshire farmer's saying has been echoed by generations of hunting people, for hound work is so intensely fascinating that it has continued long after the need to hunt for food has vanished.

In the United Kingdom over a quarter of a million people enjoying hunting,either mounted or on foot. Some hunt to ride, others for the social occasion, but the real countryman-at-heart is concerned chiefly with watching hounds work. Through their varied markings it is usually not too difficult to recognize some individual hounds (the huntsman, of course, knows them all) and to note how Dimple took up the line on the newly-ploughed furrows, or Diamond crossed the beck, gave tongue and brought a streaming flow of black, white and tan to her side in full cry.

The various types of hound have been systematically bred for centuries. The aims are clearly defined; drive and staying power, a desire to run with the pack, a good nose and a melodious voice.

Few sounds on earth compare with a well-bred pack of hounds giving tongue, and of all hounds, the Bassett is the most melodious. These short-legged dogs are bred for hunting the hare and if they seldom catch one, what of it?

The same may be said of the Beagle, slightly faster, but bred so that an active man may keep with it. If the kill were the main objective, all hounds would look like Greyhounds.

Another hound with wonderful voice is the Bloodhound, still used on the Isle of Man in tracking humans for sport. No other hound will own to as cold a line, and stick to it. Bloodhound colours are black-and-tan and liver-and-tan, possibly flecked with white.

Foxhounds are divided into lowland and fell types. The latter are lighter in bone and colour, to combat the steep slopes and rocky climbs of hill country, and so that they can be seen against dark scree or heather. Hill men think that no Foxhound can compare with a fell pack for music, so often enhanced by the echoes of mountain passes. From time to time fell blood is introduced into lowland packs. Six packs – the Vale of Lune, Coniston, Melbreak, Ullswater, Blencathra, Eskdale and Ennerdale – hunt the English Lake District, and their commission is the control of lamb-killing foxes rather than sport. Hunting continues well into May, starting before the early dawn.

Welsh Foxhounds vary with the different packs, but tend towards hard, wiry coats of 'grizzled tan' colour.

Breeding a pack of Foxhounds has been a life-long fascination for many leading figures in public affairs. One said he would rather be Master of Foxhounds than Prime Minister of England! Though the Foxhound is more intelligent than the horse, outstanding individuals are not wanted if they will not work with the pack.

A beagle pack with huntsman

A pack of foxhounds in kennels

Foxhounds hunt mainly by scent, and if they lose the line they should 'cast' themselves by working over the ground until they hit the scent again. A good huntsman does not direct them until absolutely necessary.

The Foxhound Kennel Stud Book originated in 1908, and contains the pedigrees of registered hounds. Colours are black, tan or lemon in various combinations with white.

There are 207 packs of Foxhounds in Great Britain, with 12,200 hounds. The main breed show is at the East of England Showground, Peterborough, in July. The Peterborough Standard has come in for some criticism, on the grounds that hounds were being bred for looks, and indeed for features such as exaggerated elbows which are a distinct hindrance where speed is concerned. The ability to run 50 or 60 miles on hunting day remains the real test, and one of the joys of breeding any working dog is that the show bench cannot predominate for very long.

A whole world of art and literature revolves around the Foxhound. Artists who have painted him include Sartorius, James Pollard, the Wolstenholmes, Herring, Landseer, Ferneley, Munnings, C.

A Lakeland/Border terrier investigating possibilities. An important terrier function is to indicate whether holes are occupied

A lovely pair of working greyhounds of contrasting colours

Simpson, Lionel Edwards, Cecil Aldin, Snaffles, F. Carr, R. Furness and F. Wootton. An entire library could be stocked with hunting books. *Foxhunting* in the Lonsdale Library series lists 900 hound names. They range from the well-known Beauty, Bounty and Bowman to Dynamo, Thistle, Tauntress and Waterwitch and finally Zigzag, Zodiac and Zulu.

The Staghound, as known centuries ago, no longer exists in Britain. After World War I, hounds of all sorts were scarce, and drafts from Foxhound and Harrier packs became the basis of the modern Staghound.

Hunting the hare on horseback rather than on foot involves a different type of hound, the Harrier. Longer in the leg than the Beagle, it approaches Foxhound type more nearly than did the Old Southern type of hound. Early in the eighteenth century, foxhunting took over from harehunting, and the former slow, 'psalm-singing' Harrier of the sporting writer Nimrod's irreverent phrase went out of fashion. It would be no bad thing to see them reintroduced, now that the pleasures of hound work rather than the kill are uppermost.

The Otterhound is an attractive hound of great antiquity, its rough coat carrying a wide range of colours. Grizzle, wheaten, red, black-and-tan, blue-and-tan, black, cream and occasionally liver-and-tan are all permissible.

Terriers

Although 'there's many a song about hunting', the horse and the hound are the usual subjects. Equally indispensable is the terrier, and the various terrier breeds cover a whole range of activities.

Every working terrier's instinct is, or should be, to go to ground. It will investigate every conceivable hole. Once there, differences in character appear, and while some terriers will try to fight any quarry they find there, the terrier man usually wants a dog that will bolt a fox rather than kill it.

In hunting countries it is usual to 'stop' fox earths on the day before hounds meet in the area. This is to prevent foxes gaining sanctuary, but it is equally important to know that they are not inside when the earth is being stopped.

Terriers are indispensable for such duties. They are allowed to run on in front of their handler, and the chances are that if they go to ground in their own time, the fox will slip quietly away.

Though some run with the pack on a hunting day, most terriers are transported by a mounted terrier man, or in a car. They mark their position down a fox earth by baying, and if the quarry cannot be pushed out, the men above know where to start digging. The purpose of hunting in the Lake District is primarily to safeguard the newly born lambs from foxes, and nowhere is the terrier more highly regarded.

The working Lakeland, Border or Jack Russell Terrier is most used with hounds. The modern Fox Terrier has been modelled for the show bench, and the Yorkshire and Manchester Terriers are now seldom prized for their ratting qualities.

Norfolk and Norwich Terriers are both small, usually not more than 4.5 kg (10 lb) in weight when mature, and of hard, red coat. Although rather small for foxes, they are great rabbiters. The Norfolk is drop-eared and the Norwich prick-eared.

When bull baiting was stopped in 1835, breeders of the Staffordshire Bull Terrier turned their attention to dog fights. The present, more peaceful, role of these dogs is as companions.

The Irish is a large terrier. The Kerry Blue in Ireland is still a sporting dog, and the Dublin Sporting Terrier Society has rigidly imposed tests for working against rabbit or badger. Another Irish breed is the Glen of Imaal, usually wheaten, blue or brindle. Completing the Irish breeds is the soft-coated Wheaten, originally a farm dog used on rats, rabbits and badgers.

From Scotland come the Dandie Dinmont, Skye,

West Highland, Scottish and Cairn Terriers. All have histories originating among the rocks, cairns and burrows of their exacting country.

The Airedale is the largest terrier, while Welsh and Bedlington are very game types. The Sealyham was developed in Pembrokeshire as a working dog.

Greyhound Racing

The Greyhound is the thoroughbred of the canine world. It is also probably the oldest type of the oldest dog in the world, known to the Romans as the Vertigus or Celtic dog. Other forms include today's Whippet, Afghan Hound, Saluki, Borzoi or Siberian Wolfhound, the Scottish Deerhound and the diminutive toy Italian Greyhound.

Speed, stamina, staunchness, quickness in turning, a level of intelligence that does not decline to 'cunning', and that indefinable gift 'fire' are part of the Greyhound's make-up.

Colours now admitted include black, white, red, blue, fawn, fallow, brindle, or any of those colours broken with white. A male stands 68-71 cm (27-28 in.).

Greyhound racing is an extremely organized sport, the high point of which is still the Waterloo Cup, instituted in 1836. The Stud Book gained its centenary in 1982. Early forms of the sport passed to and fro between England and America, and the first English course using the electric hare was at Belle Vue, Manchester, in the 1920s.

The National Greyhound Racing Club is the controlling body in Britain. There are registered tracks, to each of which a number of licensed trainers are attached, with a few licensed private trainers.

Tracks are oval, and the length of the circuit is usually 360-455 metres (400-500 yards). If a longer race is planned, some of the course is covered twice. Hurdle races are popular with race goers, less so with

owners of greyhounds.

The electric hare is the quarry in Greyhound racing, and on its effective regulation depends the success of the race. If it is too far ahead, the dogs are unsighted; if too near they bunch.

A distinguishing coloured jacket with its trap number on is worn by each dog. The traps are a row of small kennels, with mesh fronts which swing upwards simultaneously at a signal from the starter, after the hare has passed. Some Greyhounds use their brains and cut in on the bends, beating others blindly chasing the hare. The camera is essential in helping to judge the many close finishes.

Whippet Racing

This is a very popular sport, either inside a stadium or outside at country fairs and specialist gatherings. Outside courses tend to be in straight lines, following a 'hare' as in Greyhound racing.

Racing Whippets differ from show types, but less so than in Greyhounds. Show Whippets are sometimes raced to strengthen the leg muscles.

Terrier Racing

Now often staged at country fairs, this is a spectacle for entertainment rather than a serious sport. The terriers follow a lure wound in by an inverted bicycle. They bark, fight, disappear through the side netting to previously marked rabbit holes, and generally have a great time.

Other Coursing Dogs

The Lurcher is a dog once well down the social scale, but now near the top. Formerly associated with

The slip at a coursing meeting. The hounds are raring to go

poachers and gypsies, the Lurcher in its many forms is the companion of those who love a really sporting dog, and has also become something of a status symbol.

It is a cross, possibly several ways, between a coursing hound and a working dog. Greyhound/Collie was among the early Lurchers; Greyhound, Whippet or Deerhound on hound, gundog or terrier are but a few of the combinations. A Longdog is a cross between two different types of coursing dog.

The Deerhound is an aristocrat among coursing dogs. Unobtrusive and gentlemanly, his colour is one of various shades of grey, or wheaten. Deerhounds were used on some estates, such as the Duke of Buccleuch's, to help capture deer alive so that they could be spread to other forests.

Trail Hounds

For Lakeland shepherds and farmers, foxhunting is work and sport combined. What could be more natural than that during the non-hunting, summer season – termed by Mr Jorrocks 'the season of our misery!' – they should wish to match their hounds?

From elementary mid-nineteenth-century competitions, hound trailing has developed into a highly organized sport. Trail hounds are of Foxhound colour, but are generally finer in bone. They often wear coats when not racing, as their hair has been clipped to tone up muscles, make bathing, skin and muscle care easier, and possibly to minimize wind resistance when running.

Their attributes are speed and nose. They do not need many brains: they are bred solely to pick up a scent and follow its trail at maximum speed over long distances.

Cumbria, the Scottish Borders, and parts of Ireland and the Pennines are hotbeds of the sport. The Hound Trailing Association has its own officers, who check puppy registrations and issue a stud book.

The start of a hound trail: the baying is deafening

Trails are staged as separate events and at shows and country fairs.

One of hound trailing's greatest charms is the backcloth of majestic fells. Another is the intense activity at the start and finish contrasting with the calm and peace of the wait between.

The trail is laid by two human 'trailers', who are taken to the 'split' or point furthest from home. Each carries a bound woollen rag soaked in a revolting mixture of aniseed, turpentine, paraffin and human urine. They run their separate ways, one aiming for the start and the other for the finish.

When the 'starting end trailer' is some 50 paces from the start, he halts. The baying hounds straining at the leash are released at a signal from the starter, and the cacophony of sounds fades.

It is replaced some 20 or 30 minutes later by a din of cans, whistles and calls, as the first returning hounds come into view. At first they are mere dots against scree and heather, but on they come, flat out after ten miles, to be rewarded with a bowlful of mysterious concoction as they pass the finishing line.

Over 300 puppies are born, inspected and marked each year. Like the thoroughbred's, the trail hound's official year starts on 1 January. When mature, it will run two or three ten-mile trails each week from 1 April until the autumn.

Training consists of encouraging a pup to hunt its owner's feet or for its food. Then it follows boots dressed in aniseed, the next step being a trail past the handler; always the goal is food or some edible reward.

Although a trail hound cannot be registered unless bred from recognized Foxhound packs or registered trail hounds, there seems no sure way of breeding a winner. Carefully thought out matings may give no better results than random selection. But against its background of fell and hill, hound trailing remains a tremendous sport.

The winner of a hound trail in the foreground. Three men with binoculars scan the distance for their hounds

4 Police Dogs

Among all the sophisticated equipment used by today's police forces, the dog remains indispensable. Its uses include patrols, searching for lost persons, and discovering drugs and stolen property.

In the early days the police force did not breed its own dogs. Those used were acquired after rescues, from owners who couldn't handle or control them, or following complaints from neighbours. With such noisy, aggressive animals the police began with an obvious disadvantage.

Today they breed some dogs themselves, and take others from a network of private breeders, who understand police requirements, and maintain continuity throughout the succeeding generations. So, up to a point, police dogs are bred specially for their work.

The German Shepherd Dog or Alsatian is most numerous. There is no doubt that the sight of an Alsatian instills fear and respect in the general public. Other breeds used are Dobermann, Rottweiler, Labrador, Giant Schnauzer, Boxer and a few big cross-breds. Labradors assist the Metropolitan police on guard duty at Buckingham Palace and as 'sniffer' dogs.

Formal training begins at nine or ten months. Basic obedience is taught, including heel work,

Agility and fitness are important for police dogs. This dog is practising the six-foot scale

coming when called, lying down on command, and retrieving. The last is very important, as a great deal of later work is based on it. They then fetch objects placed just out of sight, leading on to ones further away, and eventually at the end of a track. By constant repetition and careful training they learn to follow footsteps. 'Little and often' is vital, as a dog's nose tires.

In teaching property searches, the dog picks up anything with human scent, even though that scent may be 'cold', and learns to quarter the ground thoroughly.

Both handler and dog undergo training. The operation is basically team work, and the guarding of a handler by his dog tends to be natural. This trait is developed and controlled.

Woodland and area searches are taught, the ultimate object being to find a lost child or a criminal. Here it is vital to ensure that the dog barks or bays on a find, and does *not* go in. The handler must arrive as quickly as possible, to instruct the suspect or to command the dog.

Only in the case of Bloodhounds was a scented article used, and its use is a piece of Sherlock Holmes fiction that has stuck. Bloodhounds have been displaced by German Shepherd Dogs. The Bloodhound cannot do more than other breeds, though he is reputed to have a better nose than most. They are still used in sporting man hunts (see page 34) but German Shepherd Dogs are quicker, cheaper and hardier.

Fitness is essential. The dog is kept fit by daily exercise and training which goes over all past work. It is also taught to jump, at first by clear jumps over a low obstacle, progressing to both long jumps and high jumps. Later 'scaling' or 'scrambling' is tackled and

Opposite *Caught! Dog apprehends criminal*

Dogs and handlers of the Metropolitan Police Force on a search

A sniffer dog searching for drugs

then obstacle work such as walking the plank, climbing ladders or even jumping through hoops. The starting height on the scale is about 1.5 metres (5ft); dogs do not like the in-between heights around 1.3 metres (4ft 6in.). It is natural for most dogs to scale or scramble, but they are not designed to tip off the top like a cat. Landing from a height can be dangerous.

A recent introduction is the 'sniffer' dog, trained to seek out drugs. The Labrador is often used because it is smaller than the German Shepherd Dog, cheaper to feed, and able to work in more confined spaces. Sniffer dog handlers may well be policewomen. Training is by reward exercises associated with the scent of different drugs.

Another very specialized type of reward exercise was developed during World War II, when dogs were taught to detect land mines. Titbits of food were placed underground, and the dog encouraged to sniff them out. He was taught to lie down and point, but

This Black Labrador Retriever has sniffed out the drugs hidden on top of the telephone

Handler and search dog locating a casualty in Wardlow Dale, Peak District

Opposite A training exercise in snow organized by the Search and Rescue Dog Association. Here the casualty has just been uncovered

not on top of the object. Then the food was placed between metal plates, and an association with disturbed ground developed, as obviously any land mine, when being buried, must disturb the ground.

Mrs Griffin was honoured for her work during the London blitz, when she taught a number of dogs to seek for people buried under rubble.

Dogs in the Armed Services are trained much in the same way as police dogs, but do more patrol and guarding. An extension is the parachute dog, taught to be relaxed before a jump. The spectacular side is shown at Cruft's, where Service dogs jump through hoops of flame, carrying in their mouths torches lit at either end!

The Search and Rescue Dog Association (S.A.R.D.A.) was formed in 1972 and works in conjunction with the Police and Mountain Rescue Associations. The dogs, of many different breeds, are trained to search for accident victims in open country and under avalanche conditions. They must be able to work from the air-borne scent of the missing person, find him and indicate his whereabouts to the handler. Regular training courses are held and dogs are graded as Novice Search Dog and Search Dog; a certificate of merit is awarded to a dog finding a victim in a real search.

5 Guide Dogs

A Labrador doing a remarkable job in guiding. This handler was probably given a male because of his height

'Braille made the blind man literate – the guide dog made him mobile'. That this is now a hackneyed saying serves to emphasize the enormous value of guide dogs for blind people. Their effect has never been better put than by four members of the first class for guide dog owners in 1931. 'Not only has my dog given me glorious freedom and independence, never known since pre-war days, but delightful companionship', said Alan Caldwell. Thomas Ap Rhys wrote: 'With Folly, I do not mind walking at the fastest pace or even running with her'. G. W. Lamn negotiated an obstacle 'and went away merrily, the crowd saying what a good dog it was'. And Musgrave Frankland declared: 'A guide dog is almost equal in many ways to giving a man sight itself'.

Gone is the blind man's shuffle. With dog in hand, he throws his head back and strides out boldly. Yet despite their association with blind people in some cultures for 1,000 years, guide dogs have only half a century of history in Britain.

A doctor in Germany, temporarily leaving his war-blinded patient with his Alsatian, was so impressed by the dog's behaviour that he began experiments. Mrs Dorothy Harrison Eustis used her wealth and enthusiasm to study the German work, and started training in Switzerland.

From those beginings there are now 3,000 guide dog owners in Britain alone. They are catered for by The Guide Dogs for the Blind Association, a charity which organizes foster homes for 400 puppies, besides the breeding of replacements from 120 brood bitches and 25 stud dogs.

Labrador bitches are most frequently used as guide dogs. Bitches are less inclined to distractions than dogs, though dogs may be more suitable for tall people. All bitches are spayed and males castrated.

Other breeds used are German Shepherds, Golden Retrievers, a few large cross-breds, and Collies. A Collie from a sheep-working strain would not do, however, as it would show too much 'eye'.

Standards have improved. As well as breeding its own dogs, the Association can liase with a number of reliable breeders with stock of the right temperament. A dog that eventually proves unsuitable means a waste of funds, and in the early days the percentage was alarmingly high. Then a system of 'puppy walking' was introduced, whereby people in urban environments take over the rearing of the puppy from the age of six or eight weeks to ten or twelve months. In that time they house-train the dog, take it into as many situations as possible, including buses and railway stations, accustom it to children, pets and tradesmen, and teach basic obedience. The walker teaches the dog to stop at a pavement, and walk in the centre of the pavement slightly in front of its handler. Most other dogs are, of course, taught to walk at heel. These lessons are checked at the training centre.

Basically, the dog must understand that once its harness is on, it has to allow for its handler as an extension of itself. From dog height, it must realize that a cross bar well above it is too low for the handler.

It must judge width, guiding safely across a narrow gang plank. It must come to regard a moving vehicle within a certain range as a signal to stop, and to refuse its owner's command to go forward, while a stationary vehicle could be signal to proceed.

Only dogs that show happiness in their work are retained. When not on duty they have free-running exercise in open spaces. Their trainers have had a semblance of a blind person's life through being blindfolded in strange surroundings early in their own training.

Just as important as dog tuition is teaching the new owners. Dog and handler must be adjusted to each other temperamentally as well as on the vital question of comparative heights.

A new scheme of training dogs to aid deaf people has recently been introduced. The dog is taught to let its deaf owner know when the kettle boils, or when the doorbell or telephone rings. Compared to Guide Dogs for the Blind, the scheme is in its infancy, but it holds much promise.

A Long-Coated German Shepherd Dog under training with her new handler in a crowd. The trainer walks slightly behind and to the blind person's right

6 Dogs Overseas

In Britain it is now against the law to use dogs to haul a vehicle on the public highway. Even in the past, dogs never had an extensive draught role in Britain, but on the Continent several breeds were put to such work.

The Dutch barge dog or Keeshond had very heavy shoulders, and was used to help haul the barges. It was also a guard dog, and its habit of insinuating itself into the smallest possible space is said to have been acquired from living in confined conditions on board.

Another harness animal was the Bernese Mountain Dog, brought into service in such lightweight haulage activities as milk deliveries. Basket weavers and cheese makers of Switzerland used it as a draught dog.

New Zealand and Australian sheep farmers use dogs for tasks seldom attempted in Britain. 'Fetch your bowl' is a preliminary lesson to collecting or delivering the mail bag, or bringing the newspaper, which might be left at the roadside some considerable distance away.

Leading a horse by its halter shank is another most useful accomplishment, often taught by encouraging the pup to drag a length of cord with a bell attached. The youngster is naturally delighted with the sound made by its efforts.

Sheep sometimes become cast on their backs, especially when the fleece is heavy as shearing time approaches. Dogs are taught to pull on the neck wool sufficiently to help them get up.

A time saver that could be more universally adopted is 'Shut the gate'. This is taught by placing a piece of meat high enough on the gate for the dog to stand on its hind legs to reach it. After a few lessons the gate is fixed open so that slight pressure from the dog's paws will move it, and the lesson progresses from there with lots of praise.

In Australia dogs help in the stock yards. In Britain, their job tends to be over once the sheep are penned, but a good yard dog packs sheep into the race, runs on their backs to free a jam at the pen entrance, and uses its voice.

While the Border Collie in Britain is expected to be an all-round dog, in New Zealand there are 'heading dogs' and 'huntaways'. The former bring the sheep up to the shepherd and hold them there; huntaways are noisy, driving dogs which move sheep away from them, voice being dominant to 'eye'.

The Australian Kelpie is bred for special conditions, which include extreme heat. It is a mixture, derived from mainly Welsh and Scottish imports and Dingo or Australian Wild Dog blood. It is particularly useful for yard work. Its progenitors were probably breeds no longer common: the Smooth-haired Collie, the Welsh Black-and-Tan Sheepdog and the old Welsh Hillman. The Kelpie's coate is short, flat and fairly smooth, black, black-and-tan or a few other dark shades. Ears are pointed, adding to the alert and active appearance.

Unusual colour combinations are found in the Australian Cattle Dog. It is either blue-mottled or red-speckled, possibly with black or tan markings. It is a lithe dog, up to about 56 cm (22 in.) high, and has the Welsh Corgi characteristic of heeling cattle and of avoiding their backlash, yet in form is quite unlike the Corgi. The breed has now spread to American ranches. It makes a good Obedience competitor and a first class guard. But its main function is to move large herds of unruly cattle, which it does silently and often on its own initiative.

Among the best known working dogs are the Huskies or Eskimo dogs. Spending their entire lives out of doors among ice and snow, the Huskies are taught to work as soon as they are old enough to fight, a pastime which they start at a few months old and indulge in for the rest of their lives.

The leader is vital. Not only must he subjugate the other dogs, but he must pull the sledge steadily and smoothly; the track of a well-led sleigh teams runs in a straight line, unlike the wavering marks of a less competent one.

A husky weighs 36-45 kg (80-100 lb), and is

A Bernese Mountain Dog, used as a draught animal in Switzerland. This is a five-month-old puppy and already shows the heavy bone of the adult

weighed as routine, for the long, thick coats make assessment of fat layer impossible. They eat some 5 kg (12 lb) of meat at a feed at base, but that feed must last three days. On a trip they are fed daily.

At night the animals are chained to dog-lines. These are metal stakes driven into the ice, and the chains allow exercise but insufficient distance for fighting with neighbours.

While their meat is being prepared, all the dogs howl at once. Huskies do not bark – they howl, to signify interest in special occasions such as a sledge arriving or departing. Their howl is part of the northern wastes.

Completely different conditions are favoured by the Newfoundland. This breed is as much at home in water as on land, and is used for rescue work both in its native Canada and in Britain.

Impressively built and 63-68 kg (140-150 lb) in weight, the Newfoundland was also once used for draught. The coat is flat, coarse and dense, of an oily nature, and falls into place naturally even if brushed the wrong way. Besides the dull jet black or brown colour, white-and-black is found, and this type is known as Landseer after the Newfoundland models of the famous painter.

The St Bernard is bigger still. It was developed and trained for mountain rescue work, and in years gone by carried its little cask of brandy when on search. Monks on the Great St Bernard pass between Switzerland and Italy pioneered the work, using a smaller type of St Bernard than the modern show animal.

In the United States Bloodhounds are also used by mountain rescue teams and are sometimes transported to the accident site by helicopter and lowered in special harnesses.

Opposite *Eskimo dogs: these sled dogs are bred for their stamina*

A 'search and rescue' Bloodhound in California, awaiting the arrival of the helicopter with his handler

A Bloodhound suspended from a helicopter in a specially designed harness. Every year a large number of people go missing in the California National Parks, and are found by these dogs

7 Obedience Competitions

Most owners come into Obedience training through taking their boisterous pets to dog training classes. Their dogs are either disinclined to walk on lead, are noisy or have a rooted dislike of other canines. So along goes the owner to a class where others have the same problems, and there is no better way of solving them.

The pet owner usually simply requires the dog to walk neatly at heel, to lie down, to 'go to bed', and to come when called. When this is accomplished, the pet is better company for its owner, and is less of a danger to traffic, other dogs, people, and farm stock. Arriving at its first class screaming and trying to fight with every other dog on view, the trainee becomes, under an expert tutor, a nicely mannered, decent sort of dog in a matter of weeks.

Basic Obedience is taught at these classes, and as a result the handler becomes interested in competitive Obedience. This is a series of exercises taught purely and simply for competitions. Rather like dressage in the horse world, Obedience is concerned not so much with what a dog can do, but with the manner in which it is done.

Early exercises involve walking at heel; sitting, when the handler halts; coming when called (the recall) and retrieving an object.

At a higher level come picking out the handler's scent from among others, and picking out a given scent. Sending away involves going in a specified direction, and being under command at a distance.

Obedience competitions are divided into five classes: Beginner, Novice, A, B, and C. Class C is the top bracket, and the one most usually seen on television.

Beginner and Novice classes involve heel work on and off the lead, recall, retrieve and stay exercises. In Novice and Class A there is a temperament test, the dog standing near its handler. The judge approaches and runs his hand down the dog's back. Any sign of nastiness or backing away means failure and a large loss of points. Though simple, this test sorts out the very temperamental dogs.

Class A is quite a step beyond Novice. The number of commands allowable is drastically reduced, and a scent test is introduced, the handler's scent being applied to an article supplied by the handler and among others supplied by the judge.

In Class B, heel work is done without a lead, at different speeds. The dog must do a 'send away and drop on command', and judge's articles are used for both retrieve and scent.

Heel work is done at three speeds and with three positions in Class C. The dog is left in the stand, sit and down positions as the handler continues without pause. In both the two minutes' sit and the ten minutes' down, the handler is out of the dog's sight. There is a 'send away' to a specific spot, and the dog is dropped on reaching it.

In Class C, the scent objects are squares of handkerchief-sized cloth. The dog is given the judge's scent and has to find the cloth the judge handled from among other apparently identical ones.

A Border Collie bitch working out a scent exercise: she has picked up the article and will now return it to the handler

Retrieving in Obedience is not enough: accurate presentation counts

Working trials are an outdoor and more rugged version of these tests. The dogs jump and track, in addition to performing Obedience tests. In the later stages they tie in with police work.

Virtually every breed has been tried in Obedience. By far the most popular are Border Collies, who have had brains bred into them by generations of shepherds. Next come German Shepherds, then Labradors. Shetland sheepdogs also do well, but it has proved almost impossible to train any of the hound breeds to top-ranking Obedience. One lady did qualify an Alaskan Malemute to Class C, but such breeds bred to pull are not usually too responsive in other ways. Some Toys do very well, especially Papillions, but their very small size presents difficulties for the human.

All exercises have a finish: the dog sits on the handler's left, and in competition must reach this position with a neat, smart movement.

Obedience shows take place on most Saturdays throughout the year, and there may be up to ten at various locations throughout the country, each with several hundred entries. For every one in shows, there are probably up to 12 club members at home with their owners. Dogs and the countryside benefit greatly through the enthusiasm engendered by Obedience training. Over 500 Dog Training Clubs are registered with the Kennel Club.

Each year at 40 Championship Obedience Shows throughout the country, Obedience Certificates are awarded to the winning dogs and bitches in Championship Class C. The award of three of these Obedience Certificates to a dog makes it eligible for the title of Obedience Champion.

Though practical application seems limited, a trained Obedience dog will find your car keys missing in a field. One night at dusk we dropped a dog collar in a wood, and as darkness fell decided that the only course was an early morning search. But Audrey sent her Obedience-trained Border Collie Kyle along the track, and she retrieved the collar in a matter of seconds.

Obedience has its own magazine, *Dog Training Weekly,* published from 7 Greenwich South Street, London SE10.

8 Dogs in Show Business

Films

Generations of children have been thrilled by Lassie and Rin-Tin-Tin. Under clever direction and skilled camera work their exploits appear incredible. Yet all are a series of basic exercises that could be taught to many intelligent dogs.

Sound temperaments and attractive, even flashy markings are fundamental. Often acting dogs are males, because of their extra glamour, and ruff in the Rough Collie's case. That breed is still often referred to as the 'Lassie' Collie – yet Lassie was a male!

Hand signals are added to early voice commands, so that the handler is out of camera shot and his or her voice need not be dubbed out. All acting dogs begin with basic obedience commands: 'Come' (when called), 'Walk at heel', 'Sit', 'Lie down' and 'Stay'. Simple refinements are: 'Sit up', 'Lie on your side', 'Roll over' and 'Play dead dog'. Next the dog is taught to retrieve, carry an object, then drop it into some special receptacle such as a bucket. Other lessons are to shake the object and to lie down covering eyes with paws ('Say prayers').

More advanced and most effective is the bow to end a performance. The film dog must learn to dig on command, and to crawl. A more difficult part is to act exhausted, and to limp. In the film *Lassie Come Home*, the dog's acting was so effective that dog lovers protested about extreme cruelty. When showing Lassie in person, trainer John Rothwell always had him do the slow, dragging crawl, the next moment jumping about as full of pep as ever. Spectacular jumps through hoops and windows contrast with taking a tumble and leading a person by the hand.

Teaching the dog to sneeze, yawn, stretch and scratch the flea, are most time-consuming of all. It is best explained by the wife of Lassie's trainer: 'He'll sit there for hours at a time just watching one of his dogs, then when the dog stretches or yawns or something, he'll jump up, gently repeat a command, pet it, and give it something to eat, acting as though it's just done the most amazing thing!'

Such basic kindliness is essential. The camera would immediately highlight any cowed or suppressed feelings on the dog's part. If a dog is praised every time it stretches, and given a command, it will after a time come to connect the two, just as the sheepdog handler gives the command to go right or go left every time the young Border Collie does that when circling a flock.

Climbing ladders and standing up to pretend attacks are part of the acting dog's retinue. Another essential is barking on command. The dog is led into an exciting situation, and when it barks of its own accord is given the command and a lot of praise. It is also given a command upon which it stops, and excessively noisy household pets have been cured of barking by this means.

Dogs that can open doors or take a drink on command are an obvious asset on the set. The entire process may take training time every day for two years, at the end of which over 50 lessons have been

Dogs appearing in advertisements are not always required to act, but they must have a calm, amienable temperament. This Old English Sheep Dog appears in most of the ICI Dulux advertisements and has become tremendously popular

mastered. There is a world of difference between the film dog of Rin-Tin-Tin's calibre, and the dog that simply appears on stage as part of the set.

'Wonder dogs' appeared in the early movies, to be followed by the exciting era of Strongheart, Rin-Tin-Tin and Flash, those great German Shepherd Dogs who saved so many movie heroes and conquered villains. Rin-Tin-Tin appeared in 14 films from 1927 to 1930, and was said to have earned over $500,000. Within four years of Lassie's debut after World War II, about 250 trained dogs worked in motion pictures, and for their services Hollywood paid an estimated $250,000 yearly.

Stage and Television

The main requirements for dogs on the stage are easy obedience and friendliness to everyone. They are not required to act; they simply accompany an actor or actress, who may in fact dislike dogs, a fact he or she disguises through professional skill. These people, who include some TV personalities, take the dogs around on leads to gain some affinity with the animal.

A stage dog may simply have to lie in front of the fire; a big Irish Wolfhound is imposing and dignified in this capacity. Barking on command is useful.

Television advertising requires little training as clips are filmed over and over again until just right. However, basic obedience and a hearty appetite help!

With 'personality dogs' like Shep on the television programme *Blue Peter*, temperament is of extreme importance. Unfailing good nature and initiative are two of Shep's attributes. The BBC looked at over 300 puppies before visiting Audrey Hart's kennel. Lack of fear combined with responsiveness and a happy outlook was apparent in Shep, his kennel mates and relatives, and he was taken for a studio trial.

At that time zoom lenses were toweringly big, heavy things, alarming at dog height. Shep, instead of backing off, jumped up to the lens to say 'What's that?' and breathed upon it.

Shep quickly became a major part of *Blue Peter.* On one occasion he didn't appear, as there was no call for his services on that particular programme, and 3,000 children wrote asking what had happened to Shep. Ever after, the BBC obviated such correspondence by ensuring his presence, even for just a few seconds at the end. On his birthdays he received hundreds of cards.

Shep came from a long-lived line and this is essential. It is no use upsetting audiences by endearing them to a dog that is going to die young.

Circuses and Demonstrations

Circus performances involve impersonating humans. Dogs walk around on their hindlegs, sit on chairs in the way humans do, or are dressed up in little skirts or trousers and hats.

Still under training. This black Standard Poodle dressed as a pierrot is obviously enjoying his work

Poodles are often chosen. They can be clipped attractively, and their construction is such that they are well adapted to walking on their hindlegs.

Akin to circus work are demonstrations. Dog races, bucket races carrying water or sand, or jumping with or without dumbells are popular. One demonstrator had four differently coloured Border

Despite being a champion, this Cardigan Corgi is not too proud to enjoy giving pleasure to children

Dog acts usually entail jumping. This Border Collie carries a fancy dumb-bell as part of an agility demonstration

Collies: black-and-white, brown-and-white, blue merle, and tri-colour which in fact had a lot more white than the black dog. They were spaced apart in the down position behind three sets of hurdles. Children made choices as to who would win, and then the handler called the dogs to her on command. But the one the children loved best was the little brown bitch, taught to knock the hurdles down instead of clearing them.

A German Shepherd bitch would go over or under the jumps on whistle command. She might clear three hurdles and then go *under* the last one to a whistle that the cheering audience never heard.

The height of training is sight recognition. The cleverest dogs, usually German Shepherds, recognize the letters spelling their own name, though some are given a clue by almost silent command inaudible to the human ear. A German Shepherd belonging to Audrey could bark to give the impression of adding numbers. Audrey would say 'How many are three and five?' Topper would bark eight times, when her handler would relax or drop her eyebrows, and the bitch would stop immediately. Some dogs develop peculiar traits of their own. A Border Collie, Jason, would seize on any bucket and swing it round and round. He had never been taught. Some of his descendants four or five generations later swing their feeding bowls in the same manner; such traits may be developed by the trainer (or cursed by an owner!) Other dogs spring onto any vacated chair, are natural jumpers or enjoy climbing ladders. The more unusual the trick the better, and film makers now look for breeds other than Collies or German Shepherds, to give novelty in that way.

A show spectacular is to have a cardboard or wooden doll's house complete with dolls, and to set fire to the lot. Dogs dressed in an imitation fireman's uniform then dash in to rescue the dolls in their mouths.

A competent commentator adds greatly to such events. He must be knowledgeable about what is going on, and the likely mishaps, and be able to improvise if things go awry.

Competitive events between two or more dogs are an attraction but there is no point in taking them to extremes. One much-acclaimed event was a long jump between Hilary Verschoyle's Jason, a Border Collie, and top handler the late Bill Spencer's aggressive Danny. These dogs loathed each other. The dogs cleared an increasing length of chairs laid flat, side by side. The handlers usually agreed beforehand who would win, stopping the contest when the length became unmanageable. One night Jason muffed his clearance of 12 chairs, and so Hilary withdrew him. Jason was put on leash, and then released ready for the next event. However, the row of 13 chairs was still in position, and Jason tore towards them and cleared the lot. He had no intention of being beaten by Danny! It says much for both handlers when two active, powerful dogs of that kind of temperament can be persuaded to put on such exhibitions of verve and skill.

Addresses

British Field Sports Society,
59 Kennington Road,
London
SE1 7PZ

Guide Dogs for the Blind Association,
9-11 Park Street,
Windsor,
Berkshire

International Sheep Dog Society,
64 St Loyes Street,
Bedford
MK40 1EZ

PRO Dogs (Dogs for the Deaf),
London Road,
Aylesford,
Kent
ME20 7QH

The Animal Health Trust,
Lanwades Hall,
Kennett,
Newmarket,
Suffolk
CB8 7PN

The Blue Cross,
Animals' Hospital,
Hugh Street,
Victoria,
London
SW1V 1QQ

The Kennel Club,
1-4 Clarges Street,
Piccadilly,
London
W1Y 8AB

MAGAZINES

Country Life
Countryside
Dog Training Weekly
Dog World
Our Dogs
The Field

Index

Italic numerals refer to pages on which illustrations appear.